Historic England

Shopping Parades

Introductions to Heritage Assets

Summary

Historic England's Introductions to Heritage Assets (IHAs) are accessible, authoritative, illustrated summaries of what we know about specific types of archaeological site, building, landscape or marine asset. Typically they deal with subjects which lack such a summary. This can either be where the literature is dauntingly voluminous, or alternatively where little has been written. Most often it is the latter, and many IHAs bring understanding of site or building types which are neglected or little understood. Many of these are what might be thought of as 'new heritage', that is they date from after the Second World War.

Shopping parades are purpose-built rows of shops, often with generous residential accommodation above. They were built in large numbers, and with increasing architectural elaboration, from the mid-nineteenth century. Parades often comprised the commercial centre of suburban and dormitory communities, but were built on main thoroughfares, close to railway stations or tram or omnibus termini, where they might attract passing traffic as well as local shoppers. From the 1880s parades adopted a plethora of historicist styles: neo-Tudor, neo-Baroque, Queen Anne and a restrained neo-Georgian. The last predominated in the inter-war years, which might be regarded as the heyday of the shopping parade.

This guidance note was written by Kathryn A Morrison and edited by Paul Stamper.

It is one is of several guidance documents that can be accessed
HistoricEngland.org.uk/listing/selection-criteria/listing-selection/ihas-buildings/

First published by Historic England April 2016.
All images © Historic England unless otherwise stated.

HistoricEngland.org.uk/advice/

Front cover
Queen's Parade, Muswell Hill (London Borough of Haringey), a development of about 1897 by the developer James Edmondson. W. Martyn's shop, with canopy extended, is listed Grade II.

Contents

Introduction

Defining the term 'shopping parade' is a complex business. Most people would agree that a shopping parade, in its broadest sense, is a row of shops, but beyond that precise definitions vary. For some 'shopping parade' is equivalent to 'local shopping centre'. Thus in June 2012, as part of a strategy to revive local economies, the Department for Communities and Local Government (DCLG) published two documents: *Parades of Shops – Towards an Understanding of Performance & Prospects* (commissioned from Genecon) and *Parades to be Proud of: Strategies to Support Local Shops*. In these reports the term 'parade' was used as shorthand for neighbourhood shops and services. A parade was defined as: 'a group of 5 to 40 shops in one or more continuous row, with a mainly local customer base, containing a high number of small or micro-businesses with some multiples and symbol affiliates and is largely retail based . . . with some local services'.

The DCLG case studies included streets of shops that had grown organically and were, consequently, varied in character. Four sub-types of parade were identified, each based on location and catchment rather than physical form and appearance: 'local neighbourhood parade', 'local neighbourhood hub', 'radial parade' and 'radial destination'.

In order to assess the comparative significance of heritage assets, as a tool for managing and protecting the historic built environment, a 'shopping parade' must be defined and categorised primarily by form and function, rather than by location, catchment, or the complexion of its 'retail offer'. Thus shopping parades are specifically understood, for the purpose of the present document, to be **planned developments incorporating rows of shops (facing onto an outdoor space), with a strong degree of architectural uniformity**.

Variation occurs within the building type. The standard shopping parade comprises a terrace divided into regular units (three or more) by party walls, with shops on the ground floor and residential or office accommodation – whether for the occupants of the shop or for separate let – above. But not all shopping parades are of this standard type. Some, for example, stand just one storey high and have no secondary functions. Others incorporate additional facilities, such as a bank, theatre, cinema, library or public hall. These introduce complications that are difficult to reconcile architecturally, but which can lend interest to otherwise repetitious elevations.

By this definition, 'shopping parade' cannot be accepted as a synonym for 'local shopping centre'. A single shopping parade might form an important part of a local shopping centre, but only in housing estates, villages and the smallest suburbs does it comprise the shopping centre in its entirety. Many local shopping centres are made up of parade clusters. Furthermore, shopping parades can augment traditional town-centre high streets,

Figure 1
An unlisted parade on the edge of the central shopping area in Letchworth Garden City, Hertfordshire. Dating from 1935, this is in a style broadly similar to that adopted by the prolific developer Edward Lotery, which was replicated throughout the Home Counties.
© Ron Baxter

often occupying peripheral positions (Fig 1). The commercial centres of garden cities (such as Letchworth Garden City and Welwyn Garden City) and first-generation new towns (such as Stevenage) are composed largely of parades.

It is worth clarifying what is NOT accepted here as a shopping parade. Many rows of shops in suburban terraces in London, Birmingham and other cities developed in an *ad hoc* manner, with the conversion of parlours for commercial purposes, or by projecting shops over front gardens. Conversions and accretive developments of this nature, despite their semblance of uniformity, are excluded from this document, which focuses on purpose-built (that is planned) shopping parades. Nevertheless, it must be acknowledged that some buildings listed in the past as 'parades' were actually built as

residential terraces, and only later acquired shops, an example being Cedars Terrace, Nos. 2-26 Queenstown Road, London Borough of Lambeth, erected in a Gothic style in 1867 (listed Grade II), with single-storey shops being built over basement areas to the front from the 1880s. Lastly, parades are sometimes confused with arcades which, as covered shopping developments, are fundamentally different in form.

1 History

1.1 Proto-shopping-parades

The term 'shopping parade' did not enter general use until the early 20th century, although planned shopping developments bearing the word 'parade' in their name were built occasionally from around 1820, and more frequently from around 1870. In fact, the phenomenon of the shopping parade is sometimes considered to have started in the mid-Victorian period. At that time, rows of shops were built on a larger scale and with greater elaboration than heretofore – especially in the spreading suburbs of great cities. Nevertheless, architectural distinction cannot be accepted as an absolute criterion in defining this building type. Many shopping parades are stylistically plain, including significant examples from the mid-to-late 20th century. If these are accepted as shopping parades – as they indubitably are, by common usage – then uniform rows of shops built before the 1870s must also receive consideration. Strictly speaking, these fulfil the criteria set out above to define shopping parades.

The shopping parade thus has deep historical roots. In the medieval and early modern periods rows of shops were often created by institutions, wealthy merchants or local authorities as an extension of their town's market provision. Surviving examples like Nos. 34-51 Church Street, Tewkesbury, of about 1450 (listed Grade I), are rare, and of obvious importance. By the late 18th century it had become common for private developers to erect plain terraces with shops on the ground floor and housing for shopkeepers above. An example is a six-unit terrace with uniform shops facing the Market Place in Cheadle, built about 1800 and listed Grade II.

Rows of shops, however, were not yet called parades. As a popular name for a street, 'parade' emerged in a non-retail context. In the late 17th century the word, meaning show or procession, was applied to public spaces that were located close to barracks and used for military parades. In the 18th century the usage of 'parade' was extended to fashionable, paved residential streets in spa and seaside towns. Examples included the imposing North and South Parades in Bath, built in the early 1740s, and Marine Parade in Brighton, dating from the later 18th century. The choice of this street name indicated a venue where the well-to-do, rather than soldiers, might stroll up and down to show off (or parade) their finery. Alternative contemporary terms included 'walk' and 'promenade'.

Shopping had been an important – if sporadic – component of the fashionable promenade in seaside and spa resorts since the late 17th century, a prime example being the Pantiles in Tunbridge Wells, where the buildings – visually disparate above a unifying colonnade – are individually and selectively listed. In the early 19th century, several uniform developments combining fashionable shops with accommodation were planned in London and elsewhere. Notable examples included the Regent Street Quadrant and Oxford Circus (begun during the French Wars; completed 1825, now demolished). Further north was Thomas Cubitt's Woburn Buildings (now Woburn Walk; begun 1822, listed Grade II*), and south of the river was Nelson Road in Greenwich (1827-33 by Joseph Kay, listed Grade II). Crucially, in contrast to the Pantiles, both of these developments are listed as terraces rather than as individual addresses. Beyond London, Decimus Burton developed Calverley Promenade (now Calverley Park Crescent) in Tunbridge Wells (1829-35, Grade II*), again as a uniform composition, with shops. These high-class developments were, however, far from the norm.

Figure 2
A curving terrace of six shops built at 45-50 Market Place, Boston, Lincolnshire, by the Corporation in 1820. This is typical of rows of shops erected in the early 19th century: architecturally plain, with a narrow access road and small yards to the rear. Listed Grade II.

Throughout the early-to-mid 19th century, many unpretentious terraces were erected to meet shopping requirements, with shops on the ground floor and accommodation for the traders above. Usually the work of local builders rather than architects, the uniformity of these developments contrasted with the higgledy-piggledy nature of earlier shopping streets, and was sometimes associated with broader urban 'improvement' schemes. Although new streets were occasionally named 'parade' (for example Effra Parade, Brixton, London Borough of Lambeth, built on an estate laid out by the Westminster Freehold Land Society in 1855), there was little correspondence between this and the activity of shopping. One ostensibly precocious exception was Brunswick Parade (demolished), built in Pentonville, London Borough of Islington, about 1825. However, the name of this development commemorated a former place of promenade, rather than inaugurating new terminology for shopping developments. In reality, Brunswick Parade was a typical suburban shopping development of the time: 'of a very mean character [which has] deprived the residents of Pentonville of all pretensions to a rural place of abode'. It accommodated 'dining rooms' and a variety of trades, evidently providing a complete shopping centre for the area. Amongst its 31 units, by 1841, were a baker, a butcher, two cheesemongers and a fruiterer, two shoemakers and a linen draper. None of the traders, as one would expect at this date, were multiple retailers.

The occupancy of Brunswick Parade can be contrasted with that of a more centrally-located purpose-built row of shops, at 45-50 Market Place, Boston, Lincolnshire (1820, listed Grade II; Fig 2). This included typical high-street traders such as a china and glass dealer, a draper, a hosier and a hatter. Their upper-floor accommodation could be accessed by external rear stairs and the kitchens lay in cellars.

1.2 Mid-Victorian shopping parades, 1860-80

By the 1860s, in keeping with changing architectural fashion and the growth of middle-class consumerism, shopping terraces were becoming more ostentatious. Those in suburbs and resorts that were expanding in tandem with the railway system received the greatest embellishment. Rather than being situated in the centre of planned or existing communities, new shopping developments were increasingly positioned on main thoroughfares, close to railway stations or tram/omnibus termini, marginal locations where shops might attract passing traffic. By 1863, for example, 'Brunswick Parade', comprising a row of shops, had come into existence beside Crystal Palace Station in the London Borough of Bromley – a station much frequented by visitors to the area, as well as local commuters and other residents.

Another example of a parade located close to a railway station used by tourists is a stuccoed terrace of about 1860-5 on the south side of Terminus Road, Eastbourne, East Sussex (unlisted). Here, each unit had a canted bay window at first-floor level – such windows now being ubiquitous features of respectable front parlours, even those over shops – and a decorative cornice.

Unusual designs can be of special architectural interest. Nos 1-15 (odd) Nelson Terrace (about1860; listed Grade II) in Clifftown, Southend-on-Sea, is Italianate in style. Low, two-storey entrance bays separate higher, three-storey gabled blocks containing ground-floor shops: of particular interest here, given the early association of parades with the fashionable promenade, is a raised pavement that runs the length of the terrace. A short parade of c.1870 on the corner of Duke Street and Parade Street,

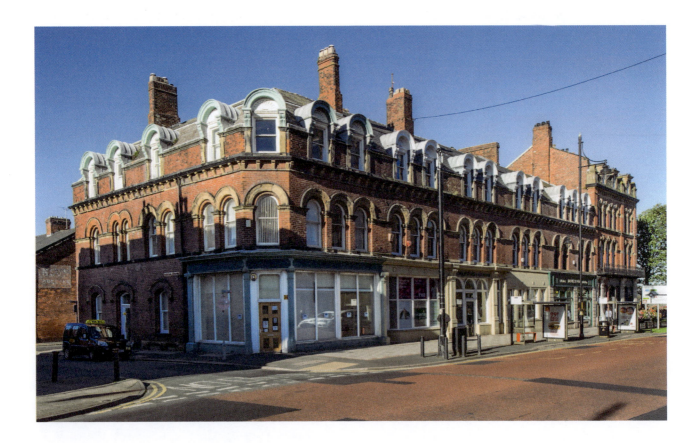

Figure 3
A Grade II-listed shopping parade of unusual design in Barrow in Furness, Cumbria, dating from about 1870.

Barrow in Furness, Cumbria, is in an attractive red brick Italianate style, with exaggerated dormers with shell tympana under projecting hoods (Fig 3). This development is enhanced by the survival of two 19th-century shopfronts, including one occupying a prominent corner position, and is listed Grade II. In a more vernacular style, Imperial Buildings, Main Street, Grange-over-Sands (of about 1870-90) is a picturesque multi-gabled development of eight shops, faced in local stone. The initial glazing scheme is near-complete, and at least one original Victorian shopfront survives intact. This is also listed, but as a 'row of shops' rather than a 'parade': perhaps a reflection of its more vernacular architectural style.

After 1870 rows of shops were increasingly called 'parade'. 'Royal Parade', Chislehurst, Kent, is a short development which displays the date 1870 (unlisted). Some years previously the colonnaded part of the Pantiles in Tunbridge Wells had been renamed 'the Parade', and the association of the word with shopping had intensified. The name 'parade', with its fashionable overtones, was presumably favoured by developers to seduce middle-class suburban residents or seaside visitors, people with aspirations to emulate their social betters, including their fondness for the shopping promenade.

In the course of the 1870s developments grew longer and higher, as illustrated by two unlisted shopping parades flanking Clapton Passage, off Lower Clapton Road in the London Borough of Hackney. The parade to the south of the Passage is plain – with restrained brick banding – but is 19 units long, with access through the centre to a rear access road. That to the north, built in 1880 and called Clapton Pavement, comprises just five units but stands three storeys high plus an attic lit by arched dormers, creating an imposing effect. As was common at this date, these parades adopted a plan form borrowed from standard terraces, comprising an unbroken series of mirror-image pairs of houses with narrow rear wings.

1.3 Late Victorian shopping parades, 1880-1900

By 1880 the fully-fledged shopping parade, with capacious upper-floor accommodation, had arrived, but humble examples continued to be built. In the 1880s numerous terraces were built in London suburbs with single-storey shops projecting from the ground floor, where one would normally expect to find a small front garden or a basement area. Indeed, this form undoubtedly originated by adding shops to the fronts of existing non-commercial buildings. A purpose-built example is Hartington Terrace (unlisted) on Queenstown Road, Battersea, London Borough of Wandsworth, erected in 1885 by the local developer Walter Peacock. The simplest way to test whether such arrangements were planned or accretive is to examine the pilasters and consoles separating the shops: if they are identical, the shops were probably built together as part of a terrace.

A very common treatment for the elevations of suburban parades through the last decades of the 19th century set a gable or a gabled dormer over an oriel or bay window. It was as if a typical terraced house was perched over each shop. Examples of this approach can be found throughout the country, with endless variants regarding materials and decoration. They can be seen in the suburbs of Birmingham, for example in Perry Barr (Birchfield Road), Sparkhill (Stratford Road) and Erdington (Station Road), and also in expanding seaside towns, such at Eastbourne, East Sussex (Grove Road, Fig 4) and St Anne's-on-Sea, Lancashire (North Crescent).

A number of centres in north London acquired clusters of parades in the late 19th and early 20th centuries. Amongst the most striking and ambitious were the builder James Edmondson's developments in Highbury Park, London Borough of Islington (1894), Crouch End, London Borough of Haringey (1895-7) and Muswell Hill, London Borough of Haringey (of about 1897; see Cover). The 42-unit Topsfield Parade in Crouch End (unlisted, Fig 5), wrapped around two sides of a triangular site with a Hippodrome

Figure 4 (top)
Grove Road, Eastbourne, East Sussex, photographed on 24 September 1886 by Bedford Lemere. A two-storey parade with banded fishscale-pattern tile hanging, decorative mock-timber gables, and uniformity from bay to bay. Today the decorative ironwork has gone, but much of the upper-floor glazing is intact: it is unlisted but typical.

Figure 5 (bottom)
Topsfield Parade, Crouch End. A 42-unit parade developed by the builder James Edmondson in 1895-97 (unlisted). Edmondson's suburban houses were similar in design to his parades.

(originally the Queen's Opera House, designed by Tom Woolnough with Frank Matcham as consulting architect) at its centre. Each three-bay unit was identical, of red brick with bright, white dressings, topped by shaped gables; just the entrance to the Hippodrome, with its entrance arch, was differentiated. The 'houses above', each designed to accommodate a single household (still mainly shopkeepers), were originally entered by stairs within the shops. Similar parades may be seen in other parts of London. Several interesting examples were built on Streatham High Road in the London Borough of Lambeth around 1890 in a Queen Anne/neo-Jacobean style with prominent corner turrets, gables, and good quality brickwork, often with terracotta dressings.

1.4 Edwardian shopping parades, 1900-14

In the early 20th century, parades adopted a plethora of styles. Neo-Tudor – with mock-framing and gables – was widely favoured, notably for parades serving new dormitory communities around London, and also throughout the West Midlands, a fine example being 35-47 Sycamore Road, Bournville (E. Bedford Tyler, 1905-8, listed Grade II; Fig 6). Other popular styles were Queen Anne, neo-Baroque and a more restrained neo-Georgian idiom. Parades were seldom stylistically pure at this time, and architects occasionally added elements from the art nouveau or arts and crafts movements to the mix.

As parades grew longer, architects became preoccupied with breaking up façades to avoid monotony, something which had clearly been

Figure 6
Nos. 37-47 Sycamore Road, Bournville, West Midlands. A well designed shopping parade, built in 1905-8 to designs by E. Bedford Tyler and listed Grade II.

an issue with many Victorian examples. This was often achieved through the technique of bay alternation, as can be seen in the design of King's Parade, built in the centre of Acton, London Borough of Ealing, in 1903. It was designed by A. H. Sykes in a heavily eclectic style and is listed Grade II. Rendered gabled bays alternated with red brick bays lit by terracotta oriels and topped by segmental-headed dormers. The first-floor windows within the rendered bays were large and elaborate, glazed in Venetian (or 'Ipswich') fashion with pargetting to the sides. As elsewhere, a number of surviving wooden consoles and pilasters show that the framework of the shopfronts – if not the glazing within the frames – was originally uniform. External light fittings were the responsibility of shopkeepers, and should be regarded as part of the shopfront.

Another way to avoid monotony was to introduce complex symmetry to either side of a central axis. The rhythm of Bennett & Bidwell's parade on Leys Avenue, in the centre of Letchworth Garden City (1908, unlisted) can be read as BABACACABAB, with A representing broad, gabled bays faced in render and B indicating narrow bays faced in red brickwork (those on the corners canted, with first-floor oculi). The blocks represented by C had level eaves but did not maintain perfect symmetry. This was typical of Bennett & Bidwell's approach elsewhere.

Secondary functions could disrupt the potentially relentless regularity of parades to good effect. An unlisted example is the small parade built by a tram stop on the edge of Aldersbrook in the London Borough of Redbridge in 1903-4. The design – executed in a neo-Tudor style, with exposed framing, tile hanging and pecked stucco – maintained symmetry to either side of a central hotel. In contrast, the hotel within the more elaborate parade at Nos.101-119 South Street, Eastbourne, of about 1900, also unlisted, contrived rather artificially to maintain the regular elevation of the units to either side, though an access road through a recessed bay split the development in two.

Occasionally, individual shops within parades have been listed at Grade II, leaving the remainder undesignated. One example is W. Martyn's, a grocer's shop in Queen's Parade, Edmondson's development in Muswell Hill (about 1897; see cover). Martyn's was listed on account of its well-preserved interior and 1930s shopfront. Another example of this approach to designation is a former butcher's shop at 157 Arthur Road, part of a nine-shop parade developed in 1904 by Ryan and Penfold to serve the Wimbledon Park Estate in the London Borough of Merton. This is listed for its original features, including the shopfront and aspects of the interior. Original shopfronts and shop interiors rarely survive from Victorian and Edwardian parades, and so these are special examples.

By 1914 the floors over parade shops were frequently leased separately, sometimes as offices or workshops, though usually as residential accommodation. The shops had effectively become lock-ups. Indeed, many were now in the hands of multiple retail organisations such as W. H. Smith, Boots or Lipton, rather than independent traders. The provision of separate entrances was crucial but problematic. Doorways on front elevations – often to the rear of lobbies shared with shop entrances – took up valuable commercial space, whilst maisonettes (that is, flats of more than one storey with their own outside entrances) with rear access commanded a lower rental. Increasingly mansion flats with a substantial shared entrance doorway on the main street elevation (and shared internal hall and stair access) were adopted as the preferred solution. The first mansion flats had been built around 1870, but only later were they incorporated with rows of shops. A mature example of this approach is Stonehills Mansion Flats and Parade, Streatham High Road, London Borough of Lambeth, (Meech & Goodall, 1905, unlisted).

Amongst the best known and most admired shopping parades of any period are the arts and crafts style Temple Fortune House and Arcade House, built on Finchley Road on the edge of Hampstead Garden Suburb (London Borough of Barnet), to designs by Arthur J. Penty, Parker & Unwin's assistant (1909-11, listed Grade II; Fig 7). These two blocks flanked the entrance

Figure 7
Arcade House and Temple Fortune House, Finchley Road, London, by Arthur J. Penty, 1909-11, listed
Grade II. These are amongst the finest shopping parades in the country.

to Hampstead Way, forming a gateway to the Suburb. The end cross-wings, with ground-floor arcading and steep half-hipped roofs (reflecting Rhenish influence), projected in front of a timber-framed row of shops with picturesque dormers and tall chimney stacks. This development was influential in the years leading up to 1914, especially for parades in north London.

Not far from Hampstead Garden Suburb, still in the London Borough of Barnet, two highly regarded parades were designed by Herbert Arthur Welch and H. Clifford Hollis (who had worked with Parker & Unwin) for opposing curved frontages on Golders Green Road (Cheapside and The Promenade, listed Grade II; Fig 8). These were longer than Penty's parades, comprising approximately 40 units each, and were consequently more difficult to design. Cheapside (about 1911-21), to the north, was saved from repetitiveness by the introduction

of mixed materials and visual accents with a strong arts and crafts flavour. The steep gables and overhanging eaves, in particular, reveal the influence of Penty's Temple Fortune parades. Welch & Hollis's earlier development, The Promenade (about 1908-9), on the south side of the street, was cast in a Wrenaissance style. While the lane behind The Promenade (Accommodation Road) was lined by mews buildings with dormers, that behind Cheapside (Golders Way) had single-storey outbuildings. In each case access to the upper-floor housing was via staircases on the rear elevation. The solution on Golders Way was more polite than the usual fire-escape-style arrangements: steep external stone stairs led to paired doorways within lean-to porches serving each group of flats.

Several parades of this period have been listed because of associated features. A triangular parade by Turner & Higgins, with

10

Figure 8
Cheapside (right; begun 1911) and The Promenade (left; begun 1908) line Golders Green Road, London.

Both are listed Grade II.

shops facing both Hendon Lane and Regents Park Road, London Borough of Barnet – and with an access lane joining the two roads to complete the triangle – accommodates King Edward Hall, a private banqueting hall, on the upper floors, rather than the usual residential accommodation (1911-12, listed Grade II). The Hall is architecturally notable, with stone oriels, wrought-iron balconies and bold lettering. This unusual parade is further distinguished by a five-storey circular clock tower on the corner.

Another parade listed purely for its architectural qualities is Nos. 188-98 Kennington Lane in the London Borough of Lambeth, by Adshead & Ramsey (1913-14, listed Grade II), a superb classical composition with intact shopfronts, developed by the Duchy of Cornwall and matching the adjacent houses.

1.5 Inter-war parades

1918-39 is often regarded as the heyday of the shopping parade. Parades were certainly built in large numbers at this time, meaning that examples for designation must be subject to a stringent selection process. The majority were erected by private developers – such as the estate agent George Cross (d.1972) who bought and developed the Manor Estate and Canons Park, both in Edgware in the London Borough of Barnet (Fig 9) – but some were built by local authorities to serve municipal cottage estates, for example by the London County Council (LCC) at Becontree and Dagenham, both in the London Borough of Barking and Dagenham, and also (in partnership with private enterprise) at Roehampton in the London Borough of Wandsworth (Fig 10). Increasingly, as the 1920s progressed, parade shops were built as lock-ups, quite divorced from

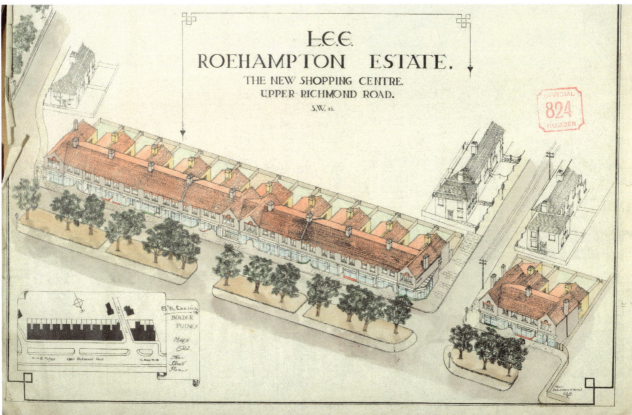

Figure 9 (top)
A typical red brick inter-war parade: the Quadrant Parade, Station Road, Edgware, London (Cowen and Cross, 1928). The speculative developer George Cross worked with various architects to create shopping parades in Edgware, Burnt Oak, Kenton, West Wickham and Cockfosters.
© Nigel Cox, Wikimedia Commons

Figure 10 (bottom)
The LCC exercised control over the layout and design of parades in several municipal housing developments. It was responsible for the neo-Georgian Dover House Parade, positioned on the edge of the Roehampton Estate and designed in 1922.
© London Metropolitan Archives (LMA/LCC/VA/DD/R245)

Figure 11
Byron Parade, Upminster, Greater London (1936), in a *moderne* style with balconies and glazed canopies. The flat roofs are dotted with chimney stacks.

the accommodation above. Inevitably, this had an impact on the layout and design.

For much of the inter-war period, a red-brick neo-Georgian style dominated. In 1928, for example, Hermitage Road in Hitchin, Hertfordshire, was lined on both sides with parades in this style (Bennett & Bidwell; unlisted), alternating parapets with strips of modillion cornice to disrupt the regularity of the scheme, and with the broad gable end of a central hall forming a strong centrepiece. Original shopfronts belonging to this scheme have transom lights with margin glazing, suggesting that shop tenants had to comply with a style guide.

Stylistic alternatives to neo-Georgian, nevertheless, abounded. King's Parade, Edgware, built for George Cross, is an example of rusticated Neo-Classicism, typical of the 1920s although its northernmost part was a later addition. More commonly, neo-vernacular (or 'stockbroker

Tudor') parades continued to be built into the early 1930s, particularly in affluent Home Counties commuter towns like Virginia Water, Surrey. A well-preserved example occupies the corner of Goring Road and Wallace Avenue in Worthing, West Sussex (built by 1932; unlisted). The baronial style – a rarer option – was chosen for the unlisted Berkeley Parade, Cranford, Middlesex, which was designed to match a nearby roadhouse called The Berkeley Arms in 1934. The outraged architect of the roadhouse, E. B. Musman, is said to have sued successfully for plagiarism. Also atypical, a Tudor (or 'modern Gothic') style was adopted for the front range of Church Close (Yates, Cook & Darbyshire, 1928, unlisted), a quadrangular development in Kensington, London Borough of Kensington and Chelsea. High-quality developments of this type tended to be compact: another example was Monmouth House, Watford, Hertfordshire (1929, by H. Colbeck), representing the partial rebuilding of a 17th-century house and for that reason listed Grade II.

By the early 1930s stepped art deco parapets were becoming ubiquitous, and fully-formed modern styles were occurring. Nos. 65-73 Streatham High Road (unlisted), for example, was built in a streamline *moderne* idiom in 1932-4 with a flat roof, a curved corner and metal windows. Byron Parade, Upminster, London Borough of Havering (1936, Fig 11) had a continuous rear access balcony, and projecting glass canopies to shelter shoppers, while The Spot in Derby (T. P. Bennett & Sons, about 1934, on local list) had a stylish *brise soleil* on the upper floor. A listed example of a modern parade, on Muswell Hill Road, London Borough of Barnet, was designed by George Coles and built on a corner site next to an Odeon Cinema in 1935. By this date roundabouts (introduced in the mid 1920s) were becoming a feature of towns and suburbs: ideal locations for shopping parades, they invited concave and convex forms that were particularly suited to a *moderne* approach, especially at the seaside, as on George V Avenue and Goring Road in Worthing (1930s, unlisted).

Regardless of style, the combination of parades with mansion flats became very common in this period (for instance, Parade Mansions, Hendon Circus, London Borough of Brent, 1920s, unlisted; Temple Fortune Parade and Mansions, Golders Green, London Borough of Barnet, 1914-24, unlisted). Some reached a great scale, with the shops effectively forming the base to monumental blocks of flats (for instance, Leigham Hall Mansions and The High, both on Streatham High Road, London Borough of Lambeth, about 1936 and 1937 respectively, both designed for the Bell Property Trust by their architects, R. Toms & Partners, unlisted). These anticipated post-war tower-and-podium arrangements.

Despite the popularity of mansion flats, many shopping parades retained rear access to each unit. Indeed, the backlands behind parades have always been worlds apart from their carefully designed commercial frontages. Typically these are untidy spaces, difficult for pedestrians or residents to negotiate. The narrow service lanes seldom have pavements. The flat roofs of shop extensions are sometimes appropriated for use by residents as rather basic patios, with plants and deckchairs. Over the years, small yards have been infilled with utilitarian stores or garages, and what remains of open space is cluttered with vehicles and rubbish bins. From the late 1920s, runs of lock-up garages on adjoining strips of land often accompanied parade developments and helped to relieve some of this chaos.

Between the wars, it was considered that a parade should include a butcher, baker, grocer, greengrocer, dairyman, newsagent/confectioner and dispensing chemist; until a district was fully developed, other trades (which now often included estate agents) were regarded as speculative. Shops typically had frontages of 18 feet to 20 feet (about 5.5 metres) and were around 50 feet (15 metres) deep. Parades, however, increasingly incorporated larger commercial units such as banks and stores. Banks normally leased corner plots and installed house-style fronts, often clad in stone or marble to convey a greater sense of permanence than neighbouring shops. Some large parade stores were occupied by independent traders. In Temple Fortune Parade, mentioned above, the southernmost premises housed C. H. Summersby's draper's shop: this was larger than the other units, rising to four storeys, with an elaborate canted end bay facing south.

The largest national multiple retailers such as Tesco, Sainsbury's and Woolworth's now exercised considerable influence, and some developed their own parades. Sainsbury's, for example, erected shopping parades through its development company, Cheyne Investments. One of its earliest developments may have been a block of four shops, with upper-floor accommodation arranged around light wells, beside Streatham Hill Station in the London borough of Lambeth (Ernest Barrow, 1920). The involvement of multiples could result in compositions punctuated with large store units. These stores often differed from standard parade shops by having upper sales floors (sometimes topped by high parapets to make up the height) rather than upper-floor residential accommodation; they often had a much deeper footprint than their neighbours, and – because multiples were sought as anchor tenants by developers – they could incorporate aspects of the retailer's preferred architectural house style. Woolworth's is known to have collaborated closely with several property developers and their agents, including Hillier, Parker, May & Rowden, Second Covent Garden Properties and Central Commercial Properties, firms which routinely employed their own staff architects to develop parades. One of the first ventures of this kind to be entered by Woolworth's was a parade on Station Road, London Borough of Harrow, designed in 1929 by North, Robin & Wilsdon, who worked hand-in-glove with Hillier, Parker May & Rowden.

By the late 1930s Woolworth's usual partner was Herman Edward Lotery (1902-87), a young entrepreneur whose company, Greater London Properties Ltd, specialised in suburban parades. As Lotery claimed: 'once a chain store has signed its lease, most of the other stores rent themselves'. Working with the agents Warwick Estates, just three firms of general contractors, and the architects Marshall & Tweedy, Lotery standardised the architectural design and construction of parades to an unprecedented degree – certainly much more than George Cross had done a decade earlier. He is estimated to have built 80 parades (equating to 1,005 shop units) around London

Figure 12
A Woolworth's store in the end unit of a symmetrical parade on the High Street, Waltham Cross, Hertfordshire, which is typical of Edward Lotery's commercial developments of the 1930s. It was completed in 1939, and photographed after the outbreak of war. Examples of this design, devised by Marshall & Tweedy, abound on the outskirts of London.

between 1930 and 1938 (when Lotery began to transfer his business activity to the USA), most of them three storey high, of dark red brick with herringbone aprons beneath the windows and a modicum of classical styling including distinctive pilaster capitals with upright foliage (Fig 12). None are listed. Lotery's house style was widely replicated, for example in Letchworth (see Fig 1) and on Jubilee Crescent, Coventry (late 1930s and 1950s, unlisted).

Another parade developer worthy of note in the 1930s was John Laing & Sons. Laing adopted a new method of accessing maisonettes over shops, something which can be seen in the company's developments at Queensbury Station Parade (1934-9), Edgware, London Borough of Brent, and Parade Mansions (about 1935, Fig 13), Watford Way, London Borough of Barnet. In each case a balcony ran at first-floor level above the shopfronts, leading to groups of two or three doorways (that is, the front doors of the maisonettes). This 'street in the sky' solution anticipated a popular post-war formula.

In the 1930s, several notable parades of lock-up shops were developed in association with London underground stations. Shops in the forecourt of

Figure 13
Parade Mansions, Watford Way, north London. Here, the upper-floor accommodation was independent of the shops and accessed by balconies on the frontage.

Brent Cross underground station, on the Northern Line in the London Borough of Barnet, (Stanley Heaps, 1923; listed Grade II) were installed within railway arches. Several examples on the southern and northern Piccadilly line extensions, by Charles Holden of Adams, Holden & Pearson, are also listed. At Acton Town (1932, Grade II), London Borough of Ealing, a simple row of single-storey shops skirts the façade of the double-height station building. At Southgate (1933), London Borough of Enfield, a more ambitious *moderne* parade (Station Parade, 1933, Grade II) curved around a circular booking hall, from which it was separated by a bus lane. The flexible shop units were two storeys high, with upper floors that could be used for storage, workrooms or flats, and could be connected with the shop or entered independently. The central part of the Southgate parade was fronted by a colonnade but, by the late 1930s, as at Upminster (see Fig 11)

it was common for parade shops to be sheltered by a continuous cantilevered canopy and/or *brise soleil*. These features were seized upon by architects, not so much because they protected shoppers, but because they provided a clear visual line between the shops and the increasingly distinct upper elevations.

1.6 Post-war parades

A new form of shopping centre, the shopping precinct (Fig 14), was introduced into England in the 1950s, based on recent European pedestrian shopping centres and on pre-war British parades. The first precincts featured in post-war new towns, in towns that were redeveloped due to extensive bomb damage, and in towns that were expanded to receive metropolitan populations displaced by bombing and slum clearance. Precincts were

Figure 14
The Stow, Harlow, Essex: a neighbourhood shopping centre designed by Frederick Gibberd and built in the 1950s, with maisonettes and offices over the 41 shops.

Originally this development admitted vehicles, but it was eventually pedestrianized. It is unlisted.
© Kathryn A Morrison

created in urban centres and, on a smaller scale, in neighbourhoods (the contemporary term for new suburbs). In the 1960s and 1970s, the precinct model was widely adopted across the country, becoming ubiquitous.

Precincts typically comprised rows of shops on the pre-war parade model. Opposing parades were often separated by continuous paving, while single parades sometimes fronted a grassed area (Fig 15). Parades normally had flats with rear access over the shops but in urban centres, such as Stevenage and Coventry, flats were commonly substituted with offices entered from the front in the manner of mansion flats. In terms of form and function, innumerable configurations were created. Some (mostly in central locations, with office accommodation) had a curtain-wall façade and flat roof, but the majority (typically in housing estates) were faced

in brick with full-frame 'punched' windows, low-pitched roofs and stumpy ridge stacks (for instance, Rectory Row Shops, East Hampstead, Berkshire, unlisted). Canopies or overhangs sheltered shoppers. In some cases the overhang was created by projecting balconies which gave access to upper-floor maisonettes (for instance, Grand Parade, Hook, Hampshire, about 1970, unlisted); others had a fully cantilevered upper storey (as at Bull Yard, Coventry, 1965 by the City Architect, Terence Gregory, unlisted, Fig 16).

While the centre of Coventry was dominated by the Upper and Lower Precincts and Bull Yard, the city suburbs acquired several ambitious variants on the precinct/parade model. At the larger end of the scale was Riley Square (1957-65), the district centre for Bell Green. Here the middle of a large plaza was occupied by a tower block, around which ranges of flats were arranged in a

Figure 15
A post-war neighbourhood shopping centre on the parade model at Willenhall Wood, Coventry, photographed in the 1960s. Typically, this faced onto a shared green space, rather than a street used by vehicles. The nursery in the foreground has been demolished and the parade altered. Unlisted.

Figure 16
Bull Yard, Coventry (1965, Terence Gregory), with the overhang of the upper storeys sheltering the ground-floor shops and a pub (The Three Tuns, which has a Grade II listed mural).

pinwheel pattern, with ground-floor shops facing onto the plaza. On a smaller scale, shops were installed beneath four blocks of 10 duplex flats on Jardine Crescent, Tile Hill (1951-5). Willenhall had a U-shaped precinct surrounded by flats with shops below (demolished), and a smaller parade at Willenhall Wood (see Fig 15).

Post-war parades generally included a mixture of delivery yards, garages and surface parking, tucked out of sight to the rear. The Staple Tye Shopping Centre serving Great Pardon in Harlow, Essex (Frederick Gibberd with Victor Hamnett, 1963-8, demolished) deployed a more expensive engineering solution to segregate pedestrians from motor vehicles. Two rows of shops were built on a deck over a service road, and the

space between them (referred to as a mall, though largely unroofed) was spanned by four blocks of maisonettes. Ramps and spiral stairs served the various levels. In the centres of large towns and cities, the fashionable tower-and-podium formula often comprised one or two levels of parade-like shops forming the podium, with flats or offices above, and a basement car park. Occasionally, the tower assumed the form of a multi-storey car park.

In the 1960s and 1970s rows of shops formed a component of many large-scale multi-functional developments, skirting the base of new hotels, theatres, office blocks and even multi-storey car parks, two of the first such examples being in Bedford (Allhallows, 1961, unlisted) and Hemel

Figure 17
Parades were often integrated with other building types after the war: here a parade of seven shops forms the street frontage of a car park in Hemel Hempstead, Hertfordshire (Fuller Hall & Foulsham, 1960, unlisted).

The mosaic map on the side elevation was by Rowland Emmett (1906-90), cartoonist and whimsical sculptor.

Hempstead, Hertfordshire (Fuller Hall & Foulsham; Hillfield/Marlowes, 1960, unlisted, Fig 17). In most of these cases, the shops were secondary to other functions, and were included to maximise the commercial potential of street frontages and subsidise other businesses.

By 1980 the precinct had been overtaken by the covered shopping mall as the preferred form for large-scale retail developments in urban centres. Elsewhere, traditional parades continued to be built, but in far fewer numbers, and with little architectural show. Recent examples include Caxton House (of about 2000), a curved development of shops and offices in Cambourne, Cambridgeshire, and a new development (2015) on Digbeth, Walsall, West Midlands. In general,

fewer local shops are created nowadays, even in large new communities such as Cambourne. Instead, much modern shopping is concentrated in edge-of-town retail parks and outlet centres where stores occupy rows of high-tech or post-modern-style warehouses which can be perceived as a single-storey shopping parades, albeit ones conceived on a titanic scale. This approach is now beginning to influence the design of urban and suburban shops, for example a small parade on The Circle, Swindon, built in 2014 to replace a small parade of 1932.

2 Change and the Future

Like the traditional corner shop, retail businesses in suburban and neighbourhood parades have been adversely affected by the decline in local shopping. This was a predictable consequence of the escalation in edge-of-town and out-of-town shopping since the 1980s, and is now exacerbated by the boom in on-line retailing. The DCLG studies cited above represent an attempt to counter this trend. Over the last 30 years the content of local parades has swung away from traditional retailing (greengrocers, butchers, small supermarkets, and so on) in favour of service industries (such as hairdressers, cafes, restaurants, takeaways, tattoo parlours, tanning salons and nail bars). By and large, parades are surviving this shift, but with such a high turnover of occupants they are increasingly unlikely to retain original shop interiors and frontages.

In many ways, whether in local shopping centres or town centres, historic parades are subject to the same pressures as terraced houses or blocks of flats. For example, their appearance might be transformed by the wholesale replacement of roofing materials or windows. In Victorian and Edwardian terrace-type parades it is common for individual units to be transformed – often by rebuilding bay windows – damaging the homogeneity of the row. A great many 20th-century parades include metal staircases, railings or balconies, required to give access to accommodation over shops, and these features can also be vulnerable to replacement over time if not well looked after.

Fewer new parades are being built outside town centres than in the past – new housing schemes are often supplied with a single one-stop retail outlet (typically a supermarket) with a small surface car park – and increasing numbers of existing parades are likely to be redeveloped as their provision exceeds demand and freeholders seek to realise the full potential of sites in prime locations.

3 Further Reading

Little has been written on the history of the shopping parade. George Cross's memoir, *Suffolk Punch*, published in 1939, provides the perspective of the inter-war parade developer. Books on suburbs sometimes include useful sections on shops, especially Alan Jackson, *Semi-Detached London* (1991), 89-96. A recent study which provides a sound chronological context for London parades is Rebecca Preston and Lesley Hoskins, 'London's Suburban Shopping Parades, 1880-1939', Report for English Heritage, March 2013 (available for consultation at the Historic England Library, Swindon). A forthcoming Historic England publication will place suburban parades within their planning context. This is Joanna Smith and Matthew Whitfield, *The English Suburb* (working title), 2018.

General books on the history of shop buildings help to place parades in the wider context of retail architecture and can be useful when evaluating individual shop units (such as a chemist's or butcher's), or even the premises of particular companies (such as Boots, Woolworths or Lipton): for this see Kathryn A Morrison, *English Shops and Shopping* (2003). Banks positioned on the corner of parades usually comply with company house style: for this see John Booker, *Temples of Mammon* (1991). Very little has been written to provide a national overview of the high-street service industries such as cafes, hairdressers, beauty salons, dry cleaners, launderettes and takeaways, which also populate parades.

Those wishing to undertake more in-depth research on individual parades might consult newspapers, historical maps, and local authority planning documents such as building approval (or regulation) plans. Information is also likely to be held by retailers – many of the large multiples maintain their own archives – and developers. Old photographs may be found in the Historic England Archive in Swindon, local libraries or record offices, and contemporary postcards showing parades can often be discovered online.

4 Acknowledgements

Author
Kathryn A Morrison

Contributors
Kathryn would like to thank the following: Ron Baxter, Jon Clarke, Rachel Forbes, Peter Guillery, Lesley Hoskins, John Minnis, Rebecca Preston, Joanna Smith and Matthew Whitfield.

Editor
Paul Stamper

Design and layout
Vincent Griffin

Images
Cover: © Historic England, Lucy Millson-Watkins (DP176261)

Figure 1: © Ron Baxter

Figure 2: © Historic England, John Minnis

Figure 3: © Historic England, Alun Bull (DP169323)

Figure 4: © Historic England (BL06649)

Figure 5: © Historic England, Lucy Millson-Watkins (DP176258)

Figure 6: © Historic England, (DP167288)

Figure 7: © Historic England, Patricia Payne (DP110937)

Figure 8: © Historic England, Lucy Millson-Watkins (DP176270)

Figure 9: © Nigel Cox, Wikimedia Commons

Figure 10: © London Metropolitan Archives (LMA/LCC/VA/DD/R245)

Figure 11: © Historic England, Lucy Millson-Watkins (DP176276)

Figure 12: © Historic England Archive (fww01_01_0643_001)

Figure 13: © Historic England, Lucy Millson-Watkins (DP176265)

Figure 14: © Kathryn A Morrison

Figure 15: © Historic England Archive, Eric de Mare, (AA98_06082)

Figure 16: © Historic England, Steven Baker (DP164659)

Figure 17: © Historic England Archive (BB95/7094)

Contact Historic England

East Midlands
2nd Floor, Windsor House
Cliftonville
Northampton NN1 5BE
Tel: 01604 735460
Email: eastmidlands@HistoricEngland.org.uk

East of England
Brooklands
24 Brooklands Avenue
Cambridge CB2 8BU
Tel: 01223 582749
Email: eastofengland@HistoricEngland.org.uk

Fort Cumberland
Fort Cumberland Road
Eastney
Portsmouth PO4 9LD
Tel: 023 9285 6704
Email: fort.cumberland@HistoricEngland.org.uk

London
1 Waterhouse Square
138-142 Holborn
London EC1N 2ST
Tel: 020 7973 3700
Email: london@HistoricEngland.org.uk

North East
Bessie Surtees House
41-44 Sandhill
Newcastle Upon Tyne
NE1 3JF
Tel: 0191 269 1255
Email: northeast@HistoricEngland.org.uk

North West
3rd Floor, Canada House
3 Chepstow Street
Manchester M1 5FW
Tel: 0161 242 1416
Email: northwest@HistoricEngland.org.uk

South East
Eastgate Court
195-205 High Street
Guildford GU1 3EH
Tel: 01483 252020
Email: southeast@HistoricEngland.org.uk

South West
29 Queen Square
Bristol BS1 4ND
Tel: 0117 975 1308
Email: southwest@HistoricEngland.org.uk

Swindon
The Engine House
Fire Fly Avenue
Swindon SN2 2EH
Tel: 01793 445050
Email: swindon@HistoricEngland.org.uk

West Midlands
The Axis
10 Holliday Street
Birmingham B1 1TG
Tel: 0121 625 6870
Email: westmidlands@HistoricEngland.org.uk

Yorkshire
37 Tanner Row
York YO1 6WP
Tel: 01904 601948
Email: yorkshire@HistoricEngland.org.uk

We are the public body that looks after England's historic environment. We champion historic places, helping people understand, value and care for them.

Please contact
guidance@HistoricEngland.org.uk
with any questions about this document.

HistoricEngland.org.uk

If you would like this document in a different format, please contact our customer services department on:

Tel: 0370 333 0607
Fax: 01793 414926
Textphone: 0800 015 0174
Email: customers@HistoricEngland.org.uk

HEAG116
Publication date: April 2016 © Historic England
Design: Historic England

Printed in Great Britain
by Amazon

56431910R00018